QUALITATIVE RESEARCH AS EMERGENT INQUIRY

REFRAMING QUALITATIVE PRACTICE IN TERMS OF COMPLEX RESPONSIVE PROCESSES

QUALITATIVE RESEARCH AS EMERGENT INQUIRY

REFRAMING QUALITATIVE PRACTICE IN TERMS OF COMPLEX RESPONSIVE PROCESSES

Sheila Keegan

EMERGENT™
PUBLICATIONS

Qualitative Research As Emergent Inquiry:
Reframing Qualitative Practice In Terms Of Complex
Responsive Processes
Written by: Sheila Keegan

Library of Congress Control Number: 2011937710

ISBN: 978-0-9842165-8-1

Copyright © 2011
Emergent Publications,
3810 N. 188th Ave, Litchfield Park, AZ 85340, USA

Printed in the United States of America

ACKNOWLEDGEMENTS

I would like to thank to Ralph Stacey and Peter Allen for reading and commenting on this essay.

ABOUT THE AUTHOR

Dr **Sheila Keegan** is a business psychologist, qualitative researcher and founding partner of Campbell Keegan Ltd. She works with private and public sector clients within the areas of change and communications, to generate new thinking and new strategies for brands, products, services, internal communications and corporate strategy. Her interests lie in the complexity sciences and exploring ways of introducing improvisational and holistic thinking into qualitative research and organizational development. Sheila is also an author, a trainer in qualitative market research, a Master Practitioner of NLP, a Fellow of the Market Research Society and a Fellow of the University of Bath, School of Management. She is a regular speaker at conferences, universities and to government and commercial bodies and acts as an advisor on cultural trends and consumer psychology to companies, ad agencies, think tanks, public service companies and the media.

CONTENTS

INTRODUCTION

Marketing and social research have become part and parcel of consumer culture. Many of us, at some time or other, have participated in surveys, been recruited for focus groups or phoned to ask our opinions on political or social issues. Finding out what customers and clients—past, present and future—want or need lies at the heart of both market and social research. Since the 1930s, a variety of research methods have been used to help public and private sector organizations develop appropriate strategies and to plan their future development. At best, research provides dispassionate and structured perspectives that enable organizations to provide customers with appropriate products and services, delivered in the most effective manner.

As an industry, commercial marketing and social research have blossomed worldwide over recent decades and continue to grow, especially in emerging markets. In 2009, the global turnover for all market research spend was US$28.9bn, of which 14% was qualitative research (Esomar Industry Report, 2009). At the same time, market research techniques have increasingly been applied outside commercial markets, especially within the public sector and within the field of organizational change.

Within this broad research arena, qualitative research in particular has changed over time as new techniques have been developed and research approaches have evolved to meet client demand. In the 1980s, focus groups were the most popular methodology. Whilst they are still a core qualitative method, ethno-methodologies, semiotics, on-line text analytics, film documentary and many alternative qualitative approaches have been developed.

Particularly relevant to this essay, the way in which contemporary qualitative research is conceptualized is also evolving—and to an extent fragmenting. This essay explores these developments; how qualitative research is now carried out within a commercial context, the influences of shifting paradigms and the importance of theoretical understanding for current practice. It draws on qualitative observation and research practice within a wide range of client companies during more than thirty years of commercial qualitative practice, as well as from recent, on-going conversations with other commercial practitioners and academics, and from current academic and practitioner literature.

Commercial qualitative research has always had an uneasy relationship with classical science (by which I mean a positivist world view). Although, as a research discipline, it is subject to scientific principles almost by default, (and in the absence of a more appropriate epistemological platform), it has refused to bow down to classical scientific method. Qualitative researchers have become adept at 'sitting on the fence' and juggling two conflicting ways of understanding reality—'realism' (which assumes there are facts to be discovered and harvested) and a variety of 'interpretivist' approaches (which essentially assume that knowledge is constructed). Over time, commercial qualitative research in the UK has moved away from a classic scientific paradigm towards a social constructionist perspective.

In recent years there has been growing interest amongst some qualitative practitioners, in developing a way of understanding commercial qualitative research which more closely reflects practice as it has evolved; to develop a new paradigm of commercial qualitative research (Ereaut, 2002; Valentine, 2002; Keegan, 2009a/b, 2008, 2006,2005, 2003). In parallel, academic research has also been developing new ways of understanding qualitative research (Alvesson & Skolberg, 2000; Shotter, 1993; Marshall, 1999). These approaches attempt to liberate qualitative thinking from the traditional scientific method. In essence, the aim is to unify scientific discipline with intuition and emotion and, in doing so, create a new paradigm which acknowledges the importance of all our facilities in the generation of holistic knowledge. However, this inevitably raises issues about quality, research rigour and ethics. Any alternative research approach still needs to be reliable and valid (Spencer *et al.*, 2003; Cassell *et al.*, 2005).

This essay explores how the concept of *emergence*, derived from complexity sciences, alongside contributions from neuroscience, can help us to conceptualise an holistic theory of commercial qualitative research—emergent inquiry—which embraces of the role of *emotion* in judgement and decision making.

The implications for commercial practitioners are highlighted. Training in analytical skills, emotional awareness and *reflection-in-action* (Schon, 1983) are necessary in order that analysis and interpretation of qualitative research become iterative processes of learning. Essentially, training in qualitative research needs to include qualitative *thinking* as much as *practice.*

13

This essay also attempts to highlight the creative potential of 'emergent inquiry'; how skills such as improvisation and 'living life as inquiry' (Marshall, 1999) are essential elements of contemporary commercial qualitative research. It also touches on the ways in which emergent inquiry may be validated.

QUALITATIVE RESEARCH WITHIN A COMMERCIAL ENVIRONMENT

Qualitative research is a broad term encompassing a spectrum of different practices, orientations and purposes. In essence, its purpose is to understand the 'how, what and why' of people's attitudes, beliefs and behaviors, rather than attempting to measure them, as quantitative research aims to do. In this essay I am primarily concerned with qualitative research within a commercial rather than an academic environment, because this is the area in which I have worked for the last thirty years.

However, 'commercial research' in itself is a confusing term in that it does not only refer to 'for profit' client organizations, but also encompass the public sector and 'not for profit' clients. The three areas in which commercial researchers typically work are:

1. Commercial research with 'for profit' organizations, e.g., airlines, supermarkets, manufacturers of food and drink, toiletries, toys, clothes etc., where they help companies to develop new products and services, packaging and advertising

2. 'Not for profit' organizations; typically government departments related to health, road safety, education etc., where they help clients to develop and implement strategies, e.g., on teenage road safety, educational initiatives, health programmes.

3. Organizational change and communications: Here researchers may work with either private or public companies to help them to re-focus their

organizational thinking and implement corporate
strategy.

The term 'commercial' in relation to qualitative research
defines a style of practice rather than the research
areas within which the practitioner works. Commercial
practitioners adopt a method of working which is faster
paced and more outcome oriented than is typical of
academic qualitative researchers. Training involves a
mix of formal teaching and apprenticeship; experienced
practitioners mentor trainee researchers, gradually
allowing them to take over elements of a research project.
It takes roughly four years before a researcher can be
considered to be fully fledged. Until fairly recently, there
was little formal training and no definitive qualifications,
although many qualitative researchers have social
science backgrounds. Because the training was skill
based, practitioners could be very competent but have
limited understanding of the theoretical assumptions that
informed their practice (Gordon, 1999:62). However, this
has changed radically in the last decade. Many qualitative
researchers are now very actively interested and informed
about the philosophical and practical underpinnings of
qualitative practice and there are established, in-depth
training courses. However, this will never remove the need
for extensive hands-on training.

The use of qualitative research is widespread within the
private and public sectors. It informs most commercial
marketing campaigns, public service communications
and much organizational change. However, it is not
publicly visible. This is because research outcomes are
commercially sensitive. Clients who pay good money
for their research projects do not want their competitors

unearthing their future strategies. As a result, commercial research is largely invisible unless you know where to look.

Commercial qualitative research can be traced back to the beginning of the 20th Century, possibly earlier. Anthropologists such as Boas, Mead, Malinowski, Bateson and Evans-Prichard developed a fieldwork method whereby observers immersed themselves within another culture to study the customs, habits, beliefs and behaviors of that society. Subsequently, the Chicago School of the 1920s established the importance of qualitative research in understanding the group life of human beings. These approaches enabled researchers to develop an understanding from the perspective of the 'researched'. You could say they represented the beginnings of a social constructionist epistemology.

Over recent decades, commercial qualitative research has thrived, due to the usefulness of its input to strategic and tactical decision-making within organizations. Meanwhile, at least within some academic circles, qualitative research has been largely ignored because of its perceived lack of 'objectivity' and scientific discipline. Arguably it is only fairly recently, nudged by a postmodern world view, that it has assumed a new found respectability within academia. However, the two spheres have very different aims. Academic research highlights research reliability and validity and has a strong methodological focus, whereas commercial research is judged largely by its usefulness to clients. Essentially it is research-based consultancy.

As a result of these different perspectives and usages, coupled with little or no cross fertilization of thinking, academic and commercial qualitative research practices have evolved independently and divergently. Key practical and philosophical areas of difference have arisen which have further divided the two camps and there is little shared understanding; they barely speak the same language. Unlike academia, commercial research is, in general, less constrained by a classical science paradigm. As a result, 'subjectivity'—demonstrated through experience, interpretation, and knowledge of market dynamics—is regarded as a strength rather than a weakness. Indeed most commercial researchers would argue from a social constructionist perspective that 'truth' is always in the eye of the beholder. Meanwhile, in much of academia, the scientific paradigm prevails and qualitative is often regarded as the poor relation to quantitative research.

What do commercial researchers do all day?

Commercial researchers are commissioned by a client organization to work with particular groups of people— typically customers or clients—to explore their needs, attitudes, beliefs, fears, behavior—and how these shape and are shaped in relation to others. The outcomes from this research are used to help clients develop relevant business or public sector strategies. Although there are many qualitative methodologies, for years focus groups have dominated the commercial arena. Increasingly approaches such as ethnography, co-creative methods, semiotics and Neuro-Linguistic Programming (NLP) are also employed. And, in the last five years, web based methodologies have mushroomed. On-line groups and communities, along with a wide range of data gathering online methods, have become commonplace. These new methodologies have provoked considerable debate within the research industry about the authenticity of consumer responses, the depth of analysis and even the definition of when qualitative becomes quantitative.

Typically projects are commissioned by a research buyer within a commercial organization or government department. He/she generally defines the research objectives, often with researcher input. The research agency then prepares a proposal outlining a suggested approach. This is discussed and agreed with the client. The research team then embarks on the fieldwork. Depending on the scale of the research, commissioning to debrief is unlikely to take longer than three months. This may come as something of a shock to academic researchers, but 'fast' research does not necessarily mean lack of intellectual rigour!

For instance, a government department is tasked with reducing teen road accidents and wants to understand how best to go about this. They commission a qualitative study. The researchers need to understand teenage road safety issues; how do teenagers behave when walking down a street? Are they alone? With friends? On mobiles? Talking? Where do they gather? How do they act when crossing the road? The researchers will suggest methodologies which will help them to understand teenage road behavior. This might involve group discussions, accompanying/observing teenagers, asking them to keep diaries of their movements, giving them cameras to take pictures of their environment. Throughout this process, the researchers will be looking for the triggers that will encourage teenagers to be more careful on the road, to see what messages they respond to, what images cause them to re-assess their behavior, how they can be nudged into a safer way of behaving. The researchers will then present their research findings to the client, pinpointing approaches, messages, visuals, sounds, moods, type of teenagers to be depicted, what they will wear, what they do, how they walk, how they interact with one another. Eventually hard hitting TV advertising aimed at teenagers, challenging them to change their behavior, will be developed. It is always difficult to be definitive about causality in complex social contexts, but, according to the government department, Transport for London, the number of under 16s killed or seriously injured on UK roads has been in gradual decline since 2001 and the 2001 road toll was about half the number of children killed or seriously injured in 1987. During this time, teen targeted road safety advertising has continued unabated—and has been deemed successful.

Commercial researchers are not wedded to particular methodologies. Method is a means to an end; the 'end' being actionable knowledge. A mix of (accelerated) grounded theory, discourse (and narrative) analyses and alethic heuristics—a kind of holistic 'rule of thumb' understanding which does not attempt to separate 'objective' from 'subjective' or understanding from explanation (Alvesson & Skolberg, 2000), is typically adopted—although commercial researchers would not necessarily use these terms. Where 'grounded theory' is employed, there is rarely the luxury of 'staying in the field until no further evidence emerges' (Goulding, 2002); hypotheses should be 'held lightly', and the point at which a 'good enough' understanding is reached can be sooner in commercial than academic research. If a researcher feels the need to develop or sharpen understanding, he or she may switch between methodologies without too much concern for the purity of the approach. Different methodologies may be adopted at different stages, especially during the early stages, where there is greater emphasis on capturing and making sense of individual perspectives on an issue. In later stages there is a shift towards making connections, highlighting patterns, creating a broader understanding, which will enable the client to make strategic decisions with confidence.

Many qualitative researchers, myself included, favour a social constructionist approach. We argue that data is never without interpretation, that as researchers we have to be the eternal sense-makers (Weick, 1995). Analysis of qualitative research data is an intensive sense-making process involving narrative analysis of transcripts, generating possible understandings, testing hypotheses with colleagues, making creative leaps, challenging our thinking, through to reflectivity, reflexivity, reinterpreting and so on. This activity is always driven by the key question, 'How can this best help our client make appropriate decisions and implement them?' In this sense, qualitative researchers are unavoidably consultants as well as researchers; research and its application cannot be separated. When it comes to debriefing research outcomes with the client team, this process of sense-making continues; shaping and developing knowledge to help achieve the stated objectives.

Projects are very diverse. We have outlined one fairly typical project above. Other projects have included working with staff and customers in a project to create a new shareholder magazine, with a major cosmetic company to develop products and advertising which are relevant and appealing, with a major government department as part of a process of implementing a new internal communications strategy, with the police to help define 'policing for the future'. Each project is different.

A common strand between commercial and academic researchers

The differences between commercial and academic research are significant and should not be underestimated. However, there are also underlying similarities which constrain the growth and full utilisation of qualitative research practice in both worlds.

Qualitative research within both arenas is—to different degrees—still constrained by a traditional scientific paradigm which struggles to make sense of and validate qualitative approaches. In some sectors of academia it seems as if too much effort goes into forcing a qualitative approach into a classical scientific paradigm, rather than trying to develop a new paradigm which makes sense of this type of knowledge. In commercial qualitative research, researchers are adept at 'sitting on the fence' and juggling two conflicting epistemological understandings—'realism' and a variety of 'interpretivist' approaches—at the same time. This can lead to some bizarre inconsistencies. For instance, clients may insist that the research sample and topic guide (areas to be covered in the interview) are slavishly adhered to (to avoid accusations of limited validity), although it is simultaneously accepted that the researcher is providing an informed 'consultancy' opinion, based on extensive training and past experience, as well as present learning. Researchers have learnt to straddle the divide.

In fact, the term 'qualitative research' in itself is misleading. It suggests a methodological focus, rather than a way of thinking and a route to understanding and decision making. Spotlighting methodology detracts from the less visible, painstaking and skilled thinking and analysis that is knowledge generation—and the purpose of commercial research. Consequently, it is easy for those unfamiliar with the rigour of qualitative practice to assume that it lacks discipline; that 'anyone can do it'. One result of this assumption has been the frequent denigration of 'focus groups' in academia and the media.

THE CHANGING WORLD OF COMMERCIAL QUALITATIVE RESEARCH

Before moving on to explore what a new qualitative research paradigm might look like, it is useful to briefly set the scene by examining some of the changes that have been taking place in the world of commercial research. Often it's easier to see patterns, trends, in retrospect. *"Ah, that's what was going on then!"* *"So that's what they meant when they said so and so"*. When we are in the midst of a period of intense change we are more likely to experience confusion than clarity as competing forces vie for supremacy.

We are currently experiencing such a time of change within the UK research industry. We are questioning what it is we do, how we do it and whether or not all of the activities we are engaged in can and should be subsumed under the 'research' and 'insight' umbrellas. But researchers do not work in isolation and they both influence and are influenced by adjacent professions; advertising, PR, client companies, government, media ... broadening out into politics, education, philosophy, global trends...and so on: A mesh of changing patterns.

Who knows where these changes will ultimately lead but, in the absence of a crystal ball, here are a few trends that I think are worth exploring. I am not attempting to be exhaustive, just to illustrate the issues and give a flavour of the current research industry climate:

- Over recent years, 'research' within client companies has increasingly been re-defined as 'insight', implying recognition that research is a means to an end, rather than an end in itself. 'Insight' suggests that research outcomes can be interrogated, developed and can act as a springboard for new ideas and new directions; that there is fluidity and open mindedness. This is an exciting shift, provided that it reflects a genuine change of mind-set within the client organization and within the researchers who work with them—not just a name change.

- Creativity waxes and wanes as a priority within organizational cultures. At present, many companies are seeding innovation within organizational thinking. For example, P&G have introduced a 'Creativity team', in recognition that innovation is the life-blood of the company. Creativity and innovation have become widespread corporate mantras. However, it is often difficult to tie creativity to the bottom line and in current hard times, it is an easy area to shelve—whilst still proclaiming its importance.

- The roles and relationships between researchers, ad agencies and clients appear to be becoming more fluid, interchangeable and interactive. Long term, this may reflect a stronger role for qualitative research in up-front strategic thinking.

- The researcher-consumer relationship is changing. Research participants used to be called 'respondents', reflecting the fact that the interviewer controlled the interaction and interviewee 'responded'. Co-creation—in which clients, consumers, researchers and other relevant parties work together to create and develop ideas has become fashionable and has grown into a discrete discipline. Nowadays research participants are often viewed as 'co-creating' the research outcomes.

They are regarded as experts in terms of their own experience and the researcher is trying to make sense of this expertise in the context of other participants' experiences, for the benefit of both consumers and clients. Ideally there is an even balance of power. Various combinations of researchers, clients and research participants jointly construct the research problems and the research outcomes.

- A variety of creative organizations have sprung up in recent years; brand agencies, creative hot shops, ideas factories and so on. Many encroach on areas that traditionally fell somewhere between the researcher agency and advertising agency. In part this may be because qualitative researchers have not sufficiently promoted the creative aspects of what they do. And, perhaps, it is because they fear that promoting the creativity of research will undermine their credibility as researchers.

- Qualitative research has expanded its methodological armory. In recent years, there has been a burgeoning interest in ethnography, NLP, discourse analysis, semiotics, Creative Workshops, Breakthrough Events (Langmaid & Andrews, 2003), online debates, Skype groups and much more. This diversity raises a number of issues, amongst them, 'When does research stop being research and become something else?', 'How do we evaluate the legitimacy of these approaches?'

- There is tentative but growing dialogue between commercial research practitioners and academic researchers. Each camp has very different priorities and even different 'languages', so dialogue is not straightforward. However, this dialogue is, I believe, essential to fuel the long term development of both the academic and commercial qualitative arenas.

- There is growing acceptance of research as 'consultancy' in which the explicit goal is to enhance client thinking and corporate strategic development, rather than just deliver a well honed research project. Some researchers would argue that this is divisive to the research industry, in that it undermines the 'purity' of research and the independence of the researcher. Can impartial research and consultancy live comfortably—or even uncomfortably—under the same umbrella? Some academics think not. (Gummesson, 2000). My own view is that marketing research is inevitably consultancy and that we should both accept and develop this role.

- Perhaps the most dramatic and arguably the most important change that has come about in the last five years is the influence of technology on the way in which we carry out qualitative research. Historically qualitative research has always been a 'people' discipline. It has focused on understanding human motivations, drives and behavior, conscious or otherwise—and as rational, emotional, contradictory as these often are. The aim is to understand human behavior in order to change or predict it. In this sense, qualitative research has been very clearly differentiated from quantitative research. In the last few years there had been a dramatic 'technologization' of qualitative research, including on-line focus groups and on-line research communities. Groups on Second Life have even been conducted. Then we have blogs and bulletin boards, used to facilitate data gathering. Perhaps the most contentious area is web surfing to gather data from social media sites. This raises many issues about ethics and privacy, as well as the efficacy of interpreting data without fully understanding the context in which it is created. How do we deal with the multiple personalities that people may use online? How do we make sense of sarcasm and irony using text analytic tools which are still relatively unsophisticated and in their infancy? In addition, much on-line research is carried out by researchers who are trained in quantitative methods, drawn to the area because of their familiarity with web-based technologies. But do they have the skills to interpret these qualitative data appropriately?

These themes reflect a shifting perception and practice of commercial qualitative research; a further move away from rigid structure and assumed objectivity, towards research as an on-going and creative process of action learning. Here the emphasis is placed on outcomes and the research methodologies act as steers rather than structures. However, such a change is not without its problems and challenges, notably the risk that reducing external structuring of research will result in less rigour and analysis within the research process itself.

What are the current challenges for commercial qualitative research?

The Association for Qualitative Research (AQR) is the professional body for commercial qualitative researchers in the UK. Recently it carried out a number of Round Table discussions to explore industry satisfactions and concerns—and future directions. Talking with a group of experienced qualitative researchers, the most striking aspect was the diversity of our views on what we actually do, how and why we do it. It was clear that we had very different perspectives on the role and theoretical underpinnings of our practice. These perspectives, in turn, feed our views on how we analyze and made sense of the qualitative work we are engaged in.

You might ask, given such diversity, how can we nurture a strong, cohesive industry? On the other hand, you could argue that is it precisely this diversity that makes the industry strong? My own view is that diversity is healthy, although it is important to clearly understand where the roots of our diversity lie—the differences in theoretical stance and how these shape our practice. In particular, if our perspective on 'research' is now broadening so we come to view it as a iterative process of learning, then the need for a more clearly defined theoretical understanding of qualitative practice becomes ever more urgent. Otherwise, if we abandon our traditional classical scientific underpinning altogether, research runs the risk of being reduced to the level of soapbox: We will lose our credibility and our point of difference from other consultants.

Conflicting paradigms within commercial qualitative research

At present, qualitative research is divided in its attempts to serve two masters. On the one hand, there are the strict research protocols and assumed objectivity rooted in classical science. On the other hand, there is the fluid, exploratory approach, which appears to be at odds with the scientific paradigm, but which many commercial practitioners would see as essential for good commercial qualitative research (Gordon, 1999: 109-114).

In the stereotypical scientific model we, as researchers, present ourselves as observers; forever watching, standing outside, becoming invisible even to ourselves. Our opinions are curtailed. We provide the illusion that we are neutral, uncontaminated, uncontaminating. The traditional market research language reflects this role. Much of it is passive, static, retrospective; 'reporting', 'consumers', 'respondents', 'findings', 'methodology', 'briefing', 'target', 'campaign' and conceals the active participation of researchers in the research outcomes. It is the language of butterfly collectors who catch, name and mount specimens. But in practice we are more like naturalists, working in a world of shifting relationships, changing perceptions, contextual understanding. We can never be outside the research situation. We can never be truly objective and nor would we want to be. There is no 'outside'. Our very presence changes the situation, the response, the nature of the inquiry. We all know this so well, it is second nature. And yet in the research situation we often we often pretend that we don't.

The scientific model, crudely interpreted, regards research as data gathering and the researcher as data gatherer. Often this leads to research being relegated to a back room function and to the under-acknowledgement of its input to decision making. (Polanyi, 1962) Furthermore, researchers may be excluded from implementation of the strategy that arises from their research, because this is perceived to compromise their 'objectivity'. Researchers who are ambivalent in their role as consultants, about using research to inform their thinking, may feel the need to curb their input and expressed opinions for fear of seeming to 'overstep the mark'. As a result, much of the most useful input for the client walks away in the researcher's 'head', at the end of the research presentation.

One very important consequence of 'setting ourselves apart' in this way, is that we cut off some of the most powerful sources of learning we have; our intuition and our 'whole-body' experience. The influential Portuguese neuroscientist, Antonio Damasio, has written very persuasively about the essential role that emotions and feelings (the conscious experience of emotions), play in decision making and how it is impossible for us to experience anything, consciously or unconsciously, without constant 'whole body' communication (Damasio, 2000). The mind/body split does not exist. It is a fiction. We all know this intuitively—and neuroscience reinforces this belief—but as researchers we spend a good part of our everyday lives denying it—presenting our arguments as if they are not informed by our emotions.

When we succumb to the myth of 'standing outside', we cut off parts of ourselves, important parts which feed our thinking and creativity. The false premise of 'objectivity' limits us and prevents us marshalling and utilizing all the resources at our disposal. It diminishes our potential.

The trends in commercial practice outlined above are heralding a move away from the empirical model of research, the 'detached observer' and 'unbiased' data gathering. Instead, qualitative research is increasingly understood in terms of 'social construction', which is highly dependent on historical and cultural context.

From a constructionist perspective, it is through the daily interactions between people in the course of social life that our versions of knowledge become established, i.e., research outcomes are a 'negotiated understanding' (Burr, 1995: 4-5), developed between the community of clients, researchers and research participants, within a wider cultural context.

By definition, social constructionism assumes that we *interpret* the world, rather than passively absorb 'reality' or 'fact'. 'Different people in different positions at different moments will live in different realities' (Shotter, 1993: 17), although within particular communities, there is likely to be considerable overlap between individuals' versions of reality. From this perspective, qualitative research can be viewed as creative processes of interpretation and iterative learning—or as the on-going construction of 'reality'.

Research based on classical scientific principles will differ from that which assumes a 'Constructionist' position. Each requires a different approach to research methodology, to the development of research outcomes and to the way in which these are shared—or developed in conjunction—with the client. Yet commercial researchers rarely talk explicitly to their clients about the assumptions they make or which of these theoretical positions they will adopt. Does this matter? Well, yes, because it is easy for misunderstandings to arise between clients and researchers if they have different understandings of how the research outcomes will be arrived at and communicated. A client who expects a structured 'traditional' approach will be unhappy if the researcher adopts an emergent or collaborative approach without prior discussion. Lack of clarity can create conflict.

A client who spends considerable time ensuring that the research spec is 'perfect', who wants a very detailed discussion guide for a group discussion, who asks for research concepts to be ranked in the group, who takes a literal view of what research participants say, is likely to be coming from an (implicitly) positivist perspective. Alternatively, a client who focuses more on the ideas that are emerging from the research, who makes connections with previous research, who is excited by the interaction amongst research participants, is probably coming from an (implicitly) constructionist perspective. It is likely that neither client is aware of the epistemological stance that they are adopting.

The traditional model of research, based on positivist principles, emphasizes precision in the setting of objectives, the sample specification and recruitment. Similarly, rigour in data collection, thorough analysis and accurate presentation of findings—which are differentiated from 'the recommendations'- is assumed. A linear model of communication operates, in which the client defines the 'problem', hands it to the researcher, who 'solves' the problem and then hands 'the solution' back to the client.

But this is quite a different scenario from much commercial qualitative research today. Subtly, alongside, and intertwined with, the positivist model, another approach has evolved. As befits a commercial industry, it has emerged to cater for changing client needs in a world where time is increasingly scarce and where business problems are complex and continually evolving. Nowadays, research objectives are often multi-layered, sometimes contradictory and may change as the project progresses. There might be a range of complementary research approaches used, some of which the client will participate in. Sometimes, the researchers may be the repositories of brand history; the one constant in companies with rapid staff turnover. They may be expected to act as researcher, consultant and lead decision maker. The boundaries between roles may create tensions—creative or otherwise.

The 'problem' of mixed paradigms which are not articulated

You may ask, 'Why is this a problem? Both empirical and constructionist approaches are valid and each may be used as appropriate or, indeed, they may be used in conjunction.' This is true, but it depends on both researchers and clients being very aware of which paradigm they are using, how they are using it—and when—and ensuring that research disciplines, appropriate to the paradigm, are employed. If this is to happen, there needs to be a much clearer understanding of the theory and disciplines associated with a social constructionist approach to qualitative research within the commercial arena. If this is lacking, then research quality may be jeopardized.

Take a caricatured example: Clients are watching a group discussion in a viewing lab, through a one way mirror. They are chatting, drinking wine. A junior client is simultaneously transcribing the interview content on her laptop. The transcript is being web-streamed to Cincinnati where a more senior client is in a meeting on an unrelated subject. She occasionally glances at her Blackberry to read the transcript. By the time the researcher has completed the group discussion, the clients behind the one way mirror in the UK have conferred with the senior client in the US and the decision to pull the advertising campaign has been made.

The introduction of viewing facilities, in which clients can observe research in action, have had a dramatic effect on the research process. There is an disturbing trend for clients to absorb the initial stages of qualitative research; a group discussion, an ethnographic study, and jump to their own conclusions without the apparent need for the researcher's detailed analysis, interpretation and structuring. The experience of *observing* or *participating* in the research *becomes* the research. At worst, videoed 'ethnography', may be edited to present the most articulate and photogenic research participants who will engage and entertain the sales team in a professional 3 minute vox pop, presented as 'research'. These are very different understandings of 'research' from that conducted using a scientific model (of whatever persuasion). This 'reductionist' approach to research, by-passing analysis and presenting research outcomes as bite sized, easily digestible nuggets, raises a number of questions, amongst them; 'What is research?', 'When does research stop being research and become something else—and what is this 'something else'?

Experienced commercial researchers would view research carried out in the way described above as grossly inadequate; it may be 'emergent' and involve 'co-creation', but it cannot be considered as rigorous or valid research because it misses out the key stages of disciplined thinking, analysis and interpretation. But, if it helps client decision making, has it not served its purpose? Is it not simply qualitative research evolving to meet current business needs for speed and egalitarian decision making?

And here is the rub. Many clients are not interested in the theory of qualitative research and are not aware of different qualitative traditions. It is not their job. In particular they may not be aware of the importance of analyzing and interpreting research data—and the time effort and thought required. They are interested in 'the answers'. Fair enough. These are struggles we as researchers must face alone because ultimately, if we cannot provide useful input to clients, we will be of no use to them. For this reason alone, we must develop ways of working more effectively.

The view I am developing in this paper is that research without rigour, reflection, analysis and interpretation undermines research expertise and the commercial qualitative research industry in the long term. The great strength of commercial qualitative research is qualitative thinking (Gordon, 1999: 297), creativity, the ability to make connections, analytic skills, listening, reflecting, reflexivity, improvisational skills, developing narratives and so on. If these skills atrophy, then the commercial qualitative research skill set will be lost. In particular, if our rigid methodological frame becomes more relaxed as we move towards a constructionist and emergent model of research, then it is especially important that the qualitative skill set outlined above is preserved. If we relax the classical scientific model of research and view research practice as, essentially, socially constructed, emergent, iterative learning, then we need to be clear. What theory supports this approach? What are the guidelines for practice? How do we legitimise this form of research and how do we train new commercial qualitative researchers in the appropriate skill set?

In my view, qualitative research—or more precisely, qualitative thinking—has never been more relevant in a business context than it is today. But, to make the most of the opportunities that are arising we—clients, agencies and researchers—together need to be more courageous. We need to look at what ways of understanding research—and what research practices—are most useful in this 'new age' and then decide whether or not we are willing to change our current practice so that we are relevant and willing participants in the 'new age'. I am also suggesting that if we do not evolve our practice to meet these requirements, then other professions may move in to fill the gap—and we will find ourselves side-lined and down-played within the business decision making process.

I am suggesting that we need to evolve a different perspective on qualitative research which develops, and makes better use of, the wide range of skills and abilities which are inherent in current practice, including the role of business consultant, but which, I believe, are under acknowledged. To do this effectively, we need to develop greater confidence and leadership; we need to integrate research and consultancy to a much greater extent. We also need an open acknowledgement that the researcher is part of, and inseparable from, the research process and outcomes. Above all, we need to develop acceptance of a more interactive, learning based approach to research itself.

Qualitative research is particularly well placed to make this shift. Qualitative researchers have already developed many of the skills, experience and ways of thinking which are appropriate to this new paradigm. However, we still cling—overtly or covertly—to the tail coats of traditional scientific method, even though science itself has moved on. I am suggesting a less circumscribed, but no less rigorous, form of research—'emergent inquiry'—as a new perspective on qualitative research; a perspective in which we view scientific method as a discipline, not a set of rules, in which knowledge can be fed from a broad pool of experience and where we acknowledge that research is, by definition, a creative and collaborative process.

DEVELOPING KNOWLEDGE IN A BRAVE NEW WORLD

So how do we develop a new research paradigm? I start from the premise that, in the current business and social climate, where communications are fast and multi-directional, where social and geographical mobility are endemic, where there are few certainties, traditional models of research are becoming less relevance and need to be re-assessed. Instead, we must begin with the 'world out there'—or rather the new ways in which we are making sense of this world—and build an understanding of research which mirrors it—and which better matches our clients' needs. And, I believe, we need a new research paradigm to make sense of this world.

Like most professions, when we begin to contemplate where we should be 'going', we generally start from where we are and examine the options. What new areas can qualitative research develop into? Do we need to develop new techniques? How do we 'take' qualitative research into the board room? How do we change the public perception of 'the focus group' as simply a sounding board? Navel gazing is an occupational hazard. In my youth, I spent ten years, on and off, working with a client on what retrospectively was clearly a hopeless cause; trying to make sherry popular with young people. Then along came alco-pops. Both sherry and alco-pops are alcoholic drinks, but that is where the similarity ends. This experience taught me that the real question is, often, 'What is happening out there in the real world and what can we contribute?', rather than "How can we persuade people to buy this product?" We could apply the same question to qualitative research today. Instead of asking, 'How do we make ourselves more appealing to our current and future clients?' maybe we need to ask, 'What is happening in the wider world and what could we offer that it needs'? The first is a packaging job. The second may entail a radical rethink of the future of commercial qualitative research.

However, we do not practice our research in a vacuum. Commercial research has a clear purpose. It has to be relevant to—and fit within—our current cultural and social environment. Before exploring how we might understand and practice research in a different way, we will need to take a broad look at the world we currently inhabit. What are the priorities, concerns and needs of today's world? And, more precisely, what types of knowledge are acceptable, relevant and constructive? I will start with a whirlwind tour of some of the key issues and themes that are shaping our lives and working environments and the way in which these themes are influencing how we think and make sense of the world. An ambitious aim, to be sure, but it is more to give a (rather biased) flavour than provide a full meal.

Think of some of the issues that have impinged upon your consciousness in the last week; that have entered your letterbox, your conversation or in-mail, that you have heard on the radio or seen on television. Probably these have included a fair smattering of war, globalization, house prices, technology, education, neuroscience, communications, global warming, famine, football, AIDS, the Olympics…and so on and on.

When we look around us, we see amazing things. We hear and touch extraordinary things. Our sight, hearing, touch, smell, taste are extended, wrapped around the world through the power of modern communications. We all lived through the collapse of the twin towers as it unfolded on television. We understood the scale of the Japanese earthquake even before those who were involved realized its full horror. We watch the moment of fertilization of sperm and ovum on our televisions. We can access the world from our palm. These things have become so commonplace that it is easy to forget that we experience the world in a way that no other age has experienced it. We have a huge body of knowledge to draw on. How do we make sense of it all?

As recently as a couple of generations ago, when our connections to the larger world were smaller and information flow limited, it made sense to emphasise structure in families and organizations, to define roles precisely; father went to work, mother looked after the home, children were seen and not heard, people knew their place, class barriers ensured they stayed there. There was a belief that absolute truth was possible and that God ruled the world. You could say there were fewer alternative perspectives on life. Of course it was never really that simple, but this was the general drift or at least the 'model' of the world that many people aspired or conformed to.

We may pine for these lost certainties, but the old order no longer seems so relevant or even feasible in a global environment in which different world views are commonplace, where individuals and societies are geographically and socially mobile, where there are few certainties and where 'change is the new constant'. Alvin Toffler, in his prophetic book 'Future Shock', published in 1970, anticipates 'the collapse of hierarchy' and 'the new adhocracy' (task force management). He predicts 'the fractured family', our attempts at 'taming technology', 'diversity of life-styles', a 'surfeit of sub-cultures' and the need for 'education in the future tense' (teaching children to deal with the future world, not the world of the past). It is a humbling experience to re-read the book and realize that, forty years on, we are, largely, living the future he predicted.

However, you could say that the world has not changed, but the way in which we make sense of it has changed. In recent years, there has been a fundamental shift from viewing the world as 'out there'—as immutable and constant—to viewing it as socially constructed, i.e. 'reality' is created in the relationship between the world and our perceptions of it—and this perception is constantly evolving. We see what we believe, as much as believing what we see. The implications of this apparently simple concept have had a radical impact on notions of 'truth', 'science', 'objectivity' and such like. And, of course, they also impact on the processes of marketing, social and organizational research and, as I discussed in the previous section, how we understand what it is that we are doing when we 'conduct research'. Let's look at a few of the current themes that have particular impact on how we now make sense of our world.

Some themes that shape our world

Many explanations have been offered to account for shifts in out understand of the world—and these have been endlessly debated, e.g., globalization, the welfare state, social mobility, immigration, breakdown of family values, wealth and prosperity, the cult of the individual, the gap between the rich and poor, and so on. But, for the purposes of this essay, I would like to focus on just a few key themes and examine what these may be contributing to our way of life and, in particular to qualitative research practice. Some of these themes are so familiar that they have become clichés; we no longer 'see' them and therefore underestimate their effect on us. As the Social Constructionist, John Shotter (1993: 11) puts it, they have become 'rationally invisible'. The themes I have chosen to focus on are by no means exhaustive but will, I hope, throw some light on later discussion:

The end of certainty: In a post-modern age it is difficult to hold on to a belief in absolutes. Social constructionism, within large sectors of the Western world—although clearly not in all parts of the world—has increasingly become the cultural 'norm'. Many of us, either explicitly or implicitly, have come to accept that we *construct* our world rather than merely observing what is 'out there'. We each experience things differently, depending on our past experiences, training, expectations, context etc., but generally we operate within broadly agreed cultural parameters. We 'fit' within our culture:

> …*instead of focusing immediately upon how individuals come to know the objects and entities in the world around them, we are becoming more interested in how people first*

develop and sustain certain ways of relating themselves to each other in their talk, and then, from within these ways of talking, make sense of their surroundings. (Shotter, 1993: 2)

Those of us with teenage children do not need reminding of this; a teenager's world view of, say, rights and responsibilities, is likely to be very differently constructed from that of her parents. Our parenting styles today reflect this shift; for instance parents are more likely to negotiate rules with their children, taking into account the different perspectives of parent and child. 'Right' and 'wrong' are less clear-cut. Authoritarian parenting is regarded as less 'appropriate'—or even less feasible—nowadays.

Interconnectedness: For three centuries Western science has successfully explained many of the workings of the universe, aided by the mathematics of Newton and Leibniz. It was essentially a clockwork world, one characterized by repetition and predictability; a linear, dependable world and a very useful model. But this is no longer enough to explain how the world works.

Most of nature, however, is nonlinear and is not easily predicted. Weather is the classic example: many components interacting in complex ways, leading to notorious unpredictability. Ecosystems, for instance, economic entities, developing embryos, and the brain– each is an example of complex dynamics that defy mathematical analysis or simulation. (Lewin, 1993: 11)

Nonlinear systems behave in quite different ways to linear systems. Small inputs can lead to dramatically large, but unpredictable, consequences. Many of us are familiar with the so-called 'butterfly effect'; a butterfly flaps its wings over the Amazon rain forest, and sets in motion events that lead to a storm over Chicago. The next time the butterfly flaps its wings, however, nothing of meteorological consequence happens. Nonlinear systems, and other ideas and terms from the Complexity Sciences, such as 'edge of chaos', 'self-organizing systems', 'strange attractors' and 'fractals', are gradually infiltrating their way of into everyday thinking and language. There is increasing acceptance that we cannot neatly compartmentalise or predict our world. The effects of one event may trigger a seemingly unrelated happening elsewhere. The best we can do is form an intention, act on it and accept that the outcome will be a consequence of circumstances and the mesh of our intentions along with those of others. If we really accept this proposition, think what it means in terms of the five year plan!

These notions of interconnectedness, which represent the forefront of modern science, are remarkably similar to insights from Zen Buddhism and the Vedanta philosophy of Hinduism, which were brilliantly explored back in the '60s by the theologist and writer, Alan Watts (1954, 1969). Watts conveyed the essence of Zen Buddhism through converting it into concepts which are accessible to the West. It is ironic that, at a time when, arguably, East and West are ideologically more distant than they have ever been, philosophically they are increasingly coming into alignment.

Speed: Faster and faster communications demand faster and faster response times and an expectation of 'availability'. How does this affect decision making, given that decisions based on 'all the facts' are often obsolete before they are made? Increasingly decisions have to be made, not on the basis of considered 'fact', but on predictions, anticipations of future 'fact'. The ability and willingness to make 'appropriate' decisions 'on the hoof'— or to decide not to make them—is a key requirement of leadership nowadays. In 1970, Alvin Toffler anticipated this dilemma when he said, "Every society faces not merely a succession of *probably* futures, but an array of *possible* futures, and a conflict over *preferable* futures. (Toffler, 1979: 415)

We cannot wait to be absolutely sure about the effects of global warming. We have to act 'as if' it is true. But acting 'as if' a process or event will happen does, in itself, influence the likelihood of its occurrence. Nothing can be viewed in isolation. You could say that it was ever thus; we all—from individuals to large corporations—attempt to plan our futures in one way or another and have to make decisions based on insufficient information. Perhaps the difference is that 'good enough' decision making—with its strong emphasis on speed and improvisation—is becoming more acceptable within scientific thinking, where 'objectivity', caution and fact gathering have traditionally been rooted.

The 'living present': There is a view, arising from thinking within the field of complex responsive processes (Stacey, 2003) and gaining wider acceptability, that all we can ever know is the present moment or 'the living present' (the term was also used by John Dewey in the '30s and Alan Watts in the '50s, with slightly different meanings).

The process perspective takes a prospective view in which the future is being perpetually created in the living present on the basis of present reconstructions of the past. In the living present, expectations of the future greatly influence present reconstructions of the past, whilst those reconstructions are affecting expectations. Time in the present, therefore, has a circular structure. It is this circular interaction between future and past in the present that is perpetually creating the future as both continuity and potential transformation at the same time. (Stacey, 2003: 10)

This perspective emphasizes the creative and interactive nature of the way in which we make sense of our experience. It understands life as process, rather than a series of static or contained events and implies continual movement and change. Knowledge itself can be understood in this way. It is not static, but constantly being recreated as we incorporate new perceptions and experiences into our shared experience, over time.

The 'death of personality': When I studied undergraduate psychology, several decades ago, my lecturers talked a lot about 'personality traits', which they described as remaining relatively constant regardless of context. This perspective grew out of the emphasis on the individual as a unit, rather than the individual as part of a community. Amongst people in general there was a view—and often there still is—that to change your mind shows a lack of moral fibre. Think how politicians are vilified for swapping parties or even re-considering views they first expressed years before.

However, within psychology, the emphasis has shifted towards more fluid notions of personality (Weick, 1995; Stacey, 2003), which start from the premise that we are essentially and unambiguously social creatures, even when we are on our own. These perspectives emphasis that we behave in different, and often contradictory, ways depending on the situation and the social group we are in. A caring neighbor to an Asian family may become a racist thug at a football match. Take Diana, Princess of Wales. She became iconic not just because she was a beautiful princess, but because she was such a wild and wonderful mix, like the rest of us; caring mother, innocent girl, adulteress, glamour queen, high handed, casual, scheming, a charity worker. She was a chameleon. This perspective sees us acting less as discrete units with a constant set of values and beliefs, but as part of a fluid network of relationships in which values and beliefs emerge and are played out in a variety of different ways depending on context.

'Personality trait' theory emphasizes constancy, whereas complex responsive processes theory emphasizes fluidity and flexibility. Crudely, we could say that the former focuses on *things* and the latter on the relationship *between* things. Neither view is right or wrong. It simply depends on what we choose to prioritize. And it is a matter of which is the most useful way of understanding for the task at hand. As we are required to become more and more adaptable and improvisational in our working and personal lives, a complex responsive processes view of personality may simply be more useful for making sense of our experience.

These cultural shifts have both informed and been informed by new theories and perspectives in the natural sciences, social sciences and management and organizational development which, in turn, affect the way in which we work and what is required of us. To a greater or lesser extent—and consciously as well as unconsciously—they are coloring our understanding of the world at large and our understanding of the organizations we work with; their policies and strategies, their marketing and internal communications.

In what ways does, or could, these strands of thinking impact on our work as qualitative researchers? What does it suggest to us about the way in which our practice is evolving and what might be expected of us by our clients now and in the future?

WHAT ARE THE IMPLICATIONS OF THIS 'WORLD VIEW' FOR QUALITATIVE RESEARCH PRACTICE?

These are huge areas of change that I have barely have touched upon. However, I want to try to make some connections between these cultural and societal trends and the commercial qualitative research industry; to start to understand what they might mean for our practice.

Research as social construction

A Social Constructionist perspective takes it as given that we construct our world, in the interaction between ourselves and our environment: This is an intensely creative process and cannot be otherwise. Clearly this does not imply that we all have totally different perceptions of the world. It is not a free for all. Other people, as well as the physical world itself, are part of our environment and this limits our options; we are constantly shaping and being shaped by the world around us, human and otherwise. In practice it means that we all perceive the world in different ways but, at least within the same culture, we do not usually have wildly different world views—not least because we share a common language which moulds the way we think. Indeed, if we all had different world views, buying a loaf of bread would be a major philosophical hurdle. We are hard-wired for a degree of conformity and cooperation.

As we discussed earlier, research can be viewed as a process of construction, rather than simple discovery, as we select what we pay attention to, how we interpret, structure and develop our knowledge in ways that make sense to us and, we hope, to our clients. We have no option but to interpret what our senses perceive, and the way in which we interpret sensory input is influenced by the prevailing world view or paradigm as well as our personal experience.

Given that we all make sense of the world slightly differently, depending on our past experience, training, knowledge etc., the greater the diversity of input into a research process, the greater the potential range of ideas that emerge. Engaging clients, customers, 'creatives', whoever else may contribute in some way, as co-creators, into the research process can only increase the scope of our knowledge and options. However, this does not necessarily mean that all contributions are equal or that the process needs to be egalitarian—there are judgements to be made about how best to manage the involvement of different parties in the process.

If we accept this perspective, then it has important implications for our research practice. For a start it assumes that knowledge is never static and that it is colored by cultural and personal agendas. Perhaps more contentiously, I would argue that the validity of a research approach must also be evaluated by socially constructed standards—which cannot be applied from 'outside'.

Researching the future: Finding the creative edge

As commercial qualitative practitioners, the research we are carrying out today has to relate to people's *future* needs, although we are drawing on the past to create knowledge which we hope will be relevant in the future. How can we prepare for a flu epidemic that may never arrive? How can we anticipate next year's fickle teenage fashions? If we cannot assume a linear relationship between the past and the future, how then do we answer the 'what if' questions? Just as Toffler talks about 'education in the future tense', so we need to deal with 'research in the future tense'. Increasingly we need to help our clients answer the questions that they do not yet know they have. How do we go about doing this? What can we learn from the complexity sciences, social constructionism and relationship psychology which will help us to anticipate the future?

'Researching the future' can, perhaps, become more tangible if we link it with our understanding of imagination and creativity. Complexity theorists define the 'edge of chaos', at its simplest level, as the point in a system where there is stability and instability at the same time, which Ralph Stacey describes as "the paradox of stability and instability at the same time, which we would describe as stable instability or unstable stability" (Stacey: 2011). Arguably, this is where creativity emerges. If this is so, if the 'edge of chaos' is where change and new thinking happens is not this where we should be working if we are concerned with innovation? If it is inspiration and ground breaking direction we seek, 'safe research' is unlikely to produce it. Instead we must encourage situations which we cannot totally control, but which also, at the same time, are contained, not chaotic; places where new thinking can emerge. And we must also understand how to tap the creativity in ourselves and to understand how we might encourage others—clients, research participants—to stimulate their creative potential as well.

Much work, carried out over the last two decades on the nature of creativity and idea generation, indicates that, not only are these essential components of successful business development, but they can also be developed through structured training (Puccio *et al.*, 2006). If we accept that knowledge is created, as much as discovered, then it is important that our abilities to encourage a creative orientation in ourselves, our clients, our research participants is finely honed.

In attempting to address these questions, qualitative research is seeing a resurgence of interest in creative research techniques. Sometimes this can be as simple as allowing research participants to set the agenda; issues may emerge which the researcher has not anticipated. Or it may involve some disruption in habitual patterns of behavior; psycho-drawing (in which research participants are encouraged to express their feelings through drawing), role play, deferring judgement, redefining 'strange' ideas into useful ones (Parnes, 1967). Currently more structured approaches such as Creative Workshops (Holmes & Keegan, 1983) and Breakthrough Events (Langmaid & Andrews, 2003) are increasingly being used to position 're-search' as 'future-search'. In these approaches, researchers work with diverse groups of people to re-define problems, generate new ideas and create alternative interpretations. These approaches explore possible futures that enable researchers and clients to develop hypotheses and make educated guesses about trends and cultural shifts.

To give an example: We have recently worked with a large financial organization to help them to redefine their business. This involved conversations with diverse groups of current and potential customers, of all ages, from different regions, backgrounds and gender, in order to explore their attitudes, needs, hopes, fears for their lives and those of their families—and not just their financial lives. Essentially, the aim was to understand what 'made them tick'. We approached the project in a variety of ways, one of which was a series of day long Breakthrough Events, in which a mix of people aged from 17-70, shared their lives and aspirations with us. Some of this involved 'Future-Search' techniques, in which they worked in syndicate groups to re-define problems (which they themselves had voiced) and explore possible solutions. We used a wide range of psychographic and enabling techniques, such as drawing and role play; much of the agenda was set by people themselves. Some of these sessions were videoed or photographed. We gathered a tremendous amount of useful outcomes from these sessions, but were particularly surprised at the value participants themselves seemed to derive from them. Many participants claimed that the day had given them the time to think about what they really wanted from their lives—time that they rarely found in 'real life'.

However, the most important aspect of the sessions was that our clients were also immersed in the Breakthrough Events. We needed active client participation if the work was to really feed into cultural change within the organization—and if it did not feed cultural change, then it would have been little more than entertainment. We also facilitated a number of half day Workshops within the organization. Stakeholders were encouraged to work with us to connect the outcomes from the research with their own perceptions, beliefs and also their role within the organization. We needed to make the research outcomes real and we needed clients to 'own' them, if they were to start changing 'hearts and minds'. This is not an easy process and it needs to be part of a wider agenda of cultural change, strongly supported by senior management.

The move from 'leader' to 'leadership'

Working life is not only about fixed, learnt skills or knowledge. Increasingly it is about acting authoritatively, making 'appropriate' or 'good enough' decisions in an uncertain situation, where there is insufficient information and too little time to gather it. Informed improvisation becomes the name of the game. As Organizational Consultant, Doug Griffin puts it, "Groups tend to recognise the leader role in those who have acquired a great spontaneity, a greater ability to deal with the unknown as it emerges from the known context." (Griffin, 2002: 204)

Do researchers see themselves as leading, as showing spontaneity and a greater ability to deal with the unknown? This is not a characteristic associated with the traditional researcher role, but I would argue that it is essential for a future in which we need to make informed decisions, strategic recommendations, creative leaps with limited time and partial resources. Leadership needs greater acknowledgement and development within a research context. But to lead, researchers need to be confident to express their opinions, to go beyond the research. We need to re-define 'research' as 'research consultancy' and get beyond the 'objectivity' stranglehold.

Challenging the 'linear' model of research

The speed of expected turnaround from research planning and fieldwork to feedback stage is accelerating; the phone call the next morning following the fieldwork the previous night—what were the 'top line' outcomes from last night's groups? There is no time for consideration. We feel we have to respond. We are the research experts, aren't we? With a faster turnaround and faster expectation of 'results', do we just need to think faster? Is this about developing the ability to assimilate, prioritize and generate knowledge more quickly so that we can make fast but appropriate responses? Or do we need to re-think the research model?

The way that research projects are traditionally structured, from briefing to fieldwork to analysis and interpretation, to structuring and then giving a presentation, means that we put ourselves in the position where, in accepting the project, *we* own the problem and have to deliver the solution; to 'give our client an answer'. Generally this means adopting a linear approach to research, i.e. the client passes the agreed brief to us, we go off and solve the research problem and, hopefully, come back with our 'findings' and pass the solution back to the client, like a fatted calf. Is this the most useful way of creating and conveying knowledge? Are there better ways in which clients can 'know' the research outcomes? When we start thinking along these lines, we begin to question this linear model of research. Qualitative research, in particular, does not fall naturally into this model. There is an inevitable tension in forcing non-linear processes, such as knowledge generation, which is messy and unpredictable, into a linear structure, in which much of the essence of knowledge can be lost in the process.

An alternative way of understanding research is as a more cooperative process, in which all the parties work together and spread the responsibility for generating knowledge. This can be a richer and more productive way of working. We start to consider how we can break down boundaries so that more interaction can occur. The importance of networks and viral development of knowledge is highlighted.

For instance, we have recently been working as part of a team with a client on repositioning a brand of up-market chocolates. We have worked together with the client and advertising agency to identify brand positionings, develop them in research, act in a planning role to steer thinking within the client company and translate the ideas into effective advertising. Of course, this requires a particular quality of relationship between all the parties involved, in order to avoid some individuals or parties feeling that their role or input is threatened or under acknowledged. Often these sensitivities result from historical precedent. The more we can change these out-dated patterns, the more this way of working will seem natural. It should be stressed that co-creation does not imply equal or similar input. If it did there would be no point. It is the diversity of experience and perspectives of everyone involved that create genuine innovation.

The researcher as integral to knowledge creation

Although I am advocating that researchers try to break down role boundaries and, where possible, work together in a team with client groups, I am not suggesting that individual contributions are unimportant. On the contrary. 'Interconnectivity' and 'relationship' are often interpreted as 'out there'; person to person, group to group, client to researcher, researcher to 'consumer'. However, co-creation can equally be understood as the researcher's 'conversation' within him or herself. It is what Ralph Stacey, from a complex responsive processes perspective, refers to as 'silent conversation', in which the individual is effectively going through the same processes 'internally' as the team may be engaged in 'externally' in conversation with one another.

Social Psychologist, George Herbert Mead (1962), refers to this internal conversation and problem resolution as the 'parliament of selves'. The analysis, interpretation and presentation of knowledge generated through research are hugely skilled activities which have taken a seasoned researcher years to hone. The reflections, connections, intuition and inspiration which go into generating knowledge and that distinguish knowledge from data— what we loosely label as 'analysis'—are the kernel of qualitative research. And this takes time. The fact that it is largely invisible in the research process has meant that it is often ignored, but without it, we would be just data gatherers.

How do we, as researchers, deal with these issues?

Fortunately qualitative research is blessed with having a well established pedigree in many of the qualities needed to contribute to this new perspective on the world. You could say that many qualitative researchers have been working in these ways for decades and that fashion has only just caught up with them. Researchers are expert at making connections, establishing relationships, working with the unknown, making creative leaps and creating meaning within the context of the client's problems, 'consumers' lives and the world at large.

But we are timid. We are timid in two ways. Firstly, we hang on to a model of research that we have outgrown, with a linear structure, hidebound by research protocols which emphasise monitoring and control, in a world that needs ideas and inspiration. Then we conform to a system that 'divides and rules', which further limits our potential: We separate 'consumers' from 'clients', our professional selves from our personal selves, 'data' from 'interpretation' 'logic' from 'emotion'. And indeed we often separate 'research' from 'creativity'. We 'allow' our clients to define the problem rather than helping them to explore and identify it as part of the research process. We are also timid in that, too often, we avoid speaking out, voicing our opinion, for fear that we will not be considered 'proper' researchers. Our own views are carefully laundered for client consumption.

Of course this isn't true all the time. When we know our clients well, when we feel comfortable and expansive, then we do not revert to this default mode. Our research is more imaginative, we make more connections, invite more input. But often we still feel on shaky ground because we lack the theoretical basis for this way of working and we are fearful that we will be seen as impostors.

How could we begin to change this? Let us now bring these themes together and look at how we might evolve a way of researching which in more inclusive and responsive to the world as we know it, to the needs of research and our clients and to an appropriate understanding of how we generate knowledge.

'EMERGENT INQUIRY': A NEW QUALITATIVE PERSPECTIVE

Making sense of new ways of practice

Emergent inquiry, as a term, has been used in academic literature to describe forms of collaborative or participative research (Walker, 2003; Frongillo *et al.*, 2003; Seel, 2006). However, I would see it as differing from participative *action* research (Kemmis & Taggart, 2000) in that it is not necessarily carried out as social practice for the benefit of the participants or the community. The purpose can be social and/or commercial. In essence, 'emergent' research is viewed as shared, ongoing, iterative learning, although there is no commonly agreed definition of the term. Power is not vested in one individual—the researcher—who is deemed to be the 'expert', but is spread across all of those involved; clients, customers, research participants and other interested parties.

Within the context of this paper, the term 'emergent inquiry' is used to describe a way of practicing which, to an extent, is intrinsic within the commercial qualitative tradition, but which is evolving to reflect a changing cultural climate and different client needs. From an emergent inquiry perspective, qualitative research is understood as a methodology in which the focus of the research is truly 'in the moment'; on the emergence of ideas, thoughts, feelings and how these develop, shape and are shaped by others (researchers, clients, research participants), 'moment to moment', as the generation of knowledge. From this perspective, research is temporal and holistic—rather than reductionist—in that emotion, feeling, intuition are as much a part of 'knowing' as intellectual understanding. It is nonetheless dependent on an eclectic and open-minded perspective and a toolkit of techniques to enable research participants to explore their responses to the research areas and generate new ideas.

This approach privileges knowledge creation, but this is not to suggest that the research disciplines of defining objectives, sample structuring, analysis and presentation are not critical (even if they evolve throughout the project). They are, in that they provide the structure and 'constraint' which allows something new to emerge, just as, in complexity theory, 'the edge of chaos' is the place between chaos and stagnation where creativity is fostered.

'Emergent inquiry' is an attempt to describe and start to develop a theoretical understanding for what many qualitative researchers are *really* doing, which is a combination of facilitation, observation, leadership, discourse analysis, critical thinking, reflectivity, reflexivity, emotional and sensory awareness, improvisation, analysis, hypothesis generation and testing, creative thinking, developing narrative and more, *at the same time*, as an on-going 'stream of consciousness'.

Developing qualitative research as emergent inquiry

The traditional scientific method—in the sense in which it is popularly understood -dispassionate, rational, replicable etc.—is very reassuring. It presents a world which is knowable, controllable and predictable and it enables us to make sense of many aspects of this world. Clearly it has benefits. However, it does not mirror the world as we experience it day by day; a world which is altogether less coherent and less predictable. This latter world is the one we must deal with as qualitative researchers and as human beings. As researchers, we are often working in the no man's land between the complexity and unpredictability of human nature and the need to analyze, structure and squeeze our data in a reductionist manner in an attempt to produce easily digestible outcomes for our clients, without compromising the essence of the inquiry. It is always a battle to shape and select themes, convey emotion and contradictions, argue rationally, somehow enable clients to both 'get inside the heads' of the research participants and also leave them with a sense of understanding, strategic direction…and so on. It's a daunting task and we can only ever partially succeed. We must do the best we can to combine richness of output with manageability, coherence and direction.

'Emergent Inquiry' attempt to provide a rationale for the way in which many qualitative researchers already practice, namely, attempting to combine scientific discipline, a social constructionist perspective and 'whole-body' iterative learning. The aim is to create an holistic, on-going understanding of the research outcomes and context which will enable our clients to make informed and appropriate decisions. In essence, we are trying is to loosen the boundaries of research, to make it more congruent with 'everyday life', whilst at the same time maintaining research rigour and a creative approach. Scientific method acts as a useful discipline—guidelines, not a set of rules. From this perspective, knowledge is

Figure 1 *Classical Research*

Figure 2 *Emergent Inquiry.*
On-going thoughts, feelings, emotions (experience)
Evaluation: shaping and being shaped by others and the
environment. Knowledge evolving over time.

'emergent'. This means that it is not fixed and immutable, but may develop and change over time, being generated from the input of clients, researchers, research participants and so on. Research, by this definition, is a creative and collaborative process. Researchers, clients, consumers, employees and others are encouraged to contribute different perspectives to the inquiry on the basis that greater diversity encourages greater creativity and more—and better—outcomes. I would regard emergent inquiry as a 'mind-set' or a way of practicing, rather than a collection of research techniques.

Theory that supports Emergent Inquiry

As theory, Emergent Inquiry has grown out of current thinking within the Complexity Sciences and Social Constructionism, as well as being fed by current commercial practice. Alongside changes in qualitative practice, new scientific understanding within the complexity sciences, particularly in relation to human interaction (Stacey, Shaw & Griffin, 2002), and neuroscience (Damasio, 2000) have been making significant contributions to our understanding of how human beings communicate, how knowledge is created and how we and make sense of the world as we experience it.

These are very broad fields and, for the purposes of this paper, two areas that are particularly relevant to an understanding of commercial qualitative research will be focused on; *emergence*, deriving from the theories of complex responsive processes (Stacey, 1996: 287) and *emotion* in relation to judgement and decision making (Damasio, 2000: 35-81). Complexity theorist, Professor Ralph Stacey describes complex responsive processes as "the process of interaction, or relating, which is itself a process of intending, choosing and acting. No-one steps outside to arrange it, operate on it or use it, for there is no simply objectified "it" (Stacey, 2000: 187). Stacey is highlighting a key element of emergence; it is not orchestrated from 'outside' because there is no 'inside' and 'outside'.

Complexity sciences and the notion of 'emergence'

Complexity sciences have thrown into question much of our thinking about how society functions and how knowledge is created (Stacey, 2003: 39-58). For instance, rather than thinking of society and culture as a collection of *things*, i.e. people, organizations, job roles, information, the emphasis, put simply, is on relationships *between* things. By definition, therefore, culture is fluid and ever changing and it is this cultural web of meaning that is our proper area of study (Chandler & Owen, 2002: 34.) In fact, complexity language has infiltrated marketing and the media to such an extent that terms such as tipping points, viral marketing, co-creation, herds, hubs and hive mind—

all complexity derived—are becoming part of everyday conversation.

Emergence is an important strand within the complexity sciences. Although it is a difficult concept and academics disagree as to the exact definition of the term, it broadly describes how larger patterns arise from local-level interactions. These patterns cannot be understood or predicted from the behavior of the lower-level interactions alone. Neither can they be understood in a linear way, e.g., as cause and effect. Emergence focuses on the *present moment* as our only point of experience and influence. However, whilst happening in the present, emergence unavoidably incorporates the past and the future. This idea is not new. It was formulated by St. Augustine around 400AD, developed by G.H. Mead at the beginning of the last century and has recently been described by Ralph Stacey as "the living present" (Stacey, 2011).

Web communications, crowd behavior and brand evolution can all be thought of in terms of emergence. Indeed, the recent Arab Spring can be viewed as emergent; a more or less spontaneous viral explosion of dissatisfaction with current governments.

This notion of emergence, of unpredictable and surprising outcomes is very familiar within the context of commercial qualitative research. The nature of qualitative practice means that, by definition, an exploration of the on-going inter-relationship between individuals, brands, services, their environment and the wider cultural context is a given. The study of relationships is at the heart of what qualitative research is all about. Qualitative inquiry, understood in terms of emergence, can be seen as a process of iterative learning which does not naturally lend itself to externally imposed rules and constraints. Essentially, it is the spontaneous flow and development of communication in the present moment—in which change in opinions, attitudes, mood and feeling may occur. However, this change is not by chance, but is constrained by the checks and balances that operate in the particular relationship or context (Shaw, 2002: 118-127). In a research situation, certain areas of exploration are off-bounds because they are below conscious awareness, they are irrelevant to the study or they are too intrusive. But equally, the researcher needs to delve beneath the mundane and obvious; to make new connections, to generate new ideas or emotional responses. In complexity terms, researchers are aiming for 'the edge of chaos'; the stage between chaos (complete lack of structure which inhibits creativity) and stagnation (where thinking is stuck). This is the stage where fresh thinking is most likely to emerge. At this point, the emphasis is on improvisation, on different ways of understanding or resolving the research issues in new ways.

Indeed, friction often arises when we attempt to impose a classic scientific framework on emergence interaction. Geoff Bayley (2006) describes this conflict very clearly in his analysis of the role of the interview discussion guide in a research project, torn as it is between acting as a structured set of topics and questions pre-agreed with the client *and* a tool for encouraging emergent discussion. The detailed, four page guide may be an appropriate tool for linear, prescribed research approaches, but it is inappropriate for a fluid, qualitative exploration which weaves its way between the needs of researcher and the interests of the research participants.

The classic scientific approach tends to treat *knowledge* as a 'thing' to be defined, organized, packaged cohesively, consistently and free of emotion. Essentially it is a *spatial* model of research, with knowledge fixed in time.

However, if we understand knowledge as emergent and holistic, i.e. involving mind, body and emotion and concerned with personal and shared experience rather than 'fact' or 'logic', then we must view knowledge as emotionally charged and constantly evolving. Similarly, the legitimization of research outcomes, which is part of the on-going creation of knowledge, must also be evolving. Knowledge, from this perspective, is *temporal,* i.e. changing over time. This does not imply that intellectual rigour is not important. Rather, it is incorporated within the on-going processes of generating knowledge; 'rigour' is in the processes of reflection, monitoring, evaluating— the qualitative skill set—which are all occurring *at the same time* during the research itself.

For example, in this truncated, work-a-day dialogue taken from a research interview, we see the on-going processes of evaluation, moderation and development which typically happen:

Participant:	*"No, I don't like that" (advertising concept)*
Researcher:	*"What is it you don't like about it?"*
Participant:	*"Well, the woman, her manner. She's too… well too..*
Researcher:	*"Too what?"*
Participant:	*"Too sexy. Too in your face."*

Researcher:	*"And what about the ad, as a whole, how does that strike you? What's happening there?*
Participant:	*"Well, she dominates it, so I don't like it?"*
Researcher:	*"And if she was different…?"*
Participant:	*"With a different woman, it would be different"*
Researcher:	*"In what way would it be different?"*

And so on….

This seemingly 'bread and butter' conversation is, in fact, the process of emergent inquiry in action. The researcher and participant between them explore the parameters which might change the meaning of the advertising concept. Each utterance brings forth a response from the other which slightly shifts the conversation. This response, in turn, elicits a further shift in direction and meaning— and so on (Mead,1962: 78). Together, the researcher and participant construct a useful way forward. This requires both researcher and participant to be improvising 'in the moment'; in Stacey's (2003: 10) 'living present'. This is quite different from a process in which a pre-prepared discussion guide is used, the researcher is asking questions prepared in the past and is not reacting to the specific situation with the particular individual. The researcher is not improvising 'in the moment'.

The differences between a spatial and temporal understanding of knowledge creation is illustrated below (This has been oversimplified, caricatured, in order to make the point). Figure 1 illustrates a classical research model in which the development of knowledge is a series of linear stages, whereas Figure 2 represents emergent inquiry is iterative and can be visualized as a spiral—although in reality it is a myriad of interconnected spirals.

This way of understanding research as Emergent Inquiry; fluid, on-going knowledge generation, is increasingly adopted by default in a business climate where there is information overload and where speed is becoming an over-riding necessity. It allows a more creative forum for researching the future (Gordon,1999: 281-301) and it is re-shaping research practice. For instance it has led to the virtual abandonment of the written report, to the 'instant debrief' after fieldwork (for better and worse) and to client interpretation of research before—or instead of—the researcher's analysis or presentation. It has encouraged 'client immersion' in research contexts and a wide variety of co-creation research approaches in which clients and consumers work together. We have yet to see whether these developments will improve the usefulness of qualitative research. My current view is that external and internal structures have to be balanced, so that we create a healthy tension between emergent and classical research principles.

Neuroscience and emotion

The second area, introduced earlier, that is contributing to commercial qualitative thinking, through challenging the traditional perceptions of our rational and emotional selves, is neuroscience. Antonio Damasio disputes the accepted wisdom that logic is at a 'higher level' than creativity and intuition and believes that these capabilities might be a more recent evolution than simple rationality. As the researcher Mark Earls remarks, gleefully, 'Creativity—not rationality—is the icing on the human evolutionary cake' (Earls, 2002: 25). Thinking creatively, according to Damasio, marks the current pinnacle of brain evolution (Damasio, 2000: 40-41). Using our 'whole body' to engage with an experience—in conjunction with other people—means harnessing our rationality, intuition, creative intelligence and physiological responses. Peter Senge (1990: 234) describes this experience as *alignment*, in which 'a resonance or synergy develops, like the 'coherent' light of a laser rather than the incoherent and scattered light of a light bulb' (Senge,1990: 234).

Qualitative researchers are very familiar with this whole body learning experience in a research context which involves a sense of letting go, relinquishing control, whilst at the same time steering the process. Richard Seel (2000), describing this apparent contradiction advises, 'Do not try to answer the question. Wait until the question answers itself.'

Although classical science often down-plays the importance of emotion and intuition in knowledge generation, in practice, we 'know' in every fibre of our bodies. 'Knowing' is not just a conscious activity. Admittedly we 'know' in different ways in our muscles and skin than we know with or conscious mind, but 'knowing' is not restricted to the brain, as the theologian and writer, Alan Watts eloquently explains:

> ...*we accept a definition of ourselves which confines the self to the source and to the limitations of conscious attention. This definition is miserably insufficient, for in fact we know how to grow brains and eyes, ears and*

fingers, hearts and bones, in just the same way that we know how to walk and breathe, talk and think—only we can't put it into words. Words are too slow and too clumsy for describing such things, and conscious attention is too narrow for keeping track of all their detail. (Watts, 1969: 138)

The separation of intellect, body and emotion, introduced by Descartes in the mid 17th century, is now largely discredited by neuroscience, but its shadow lives on. As discussed earlier, we still tend to play safe. Opinion, feeling and emotion are concealed—or at least contained. We act as if they do not exist whilst unavoidably employing them in every decision we make. But our emotions are an invaluable input to our understanding, so much so that their absence undermines reason. It is time to come out of the closet.

The neurological evidence simply suggests that selective absence of emotion is a problem. Well targeted and well-deployed emotion seems to be a support system without which the edifice of reason cannot operate properly. These results and their interpretation called into question the idea of dismissing emotion as a luxury or a nuisance or a mere evolutionary vestige. They also made it possible to view emotion as an embodiment of the logic of survival. (Damasio, 2000: 42)

It follows that qualitative research needs to acknowledge emotional experience—expressed by research participants, researchers, clients and other stake-holders— as valid input to research. Our feelings and opinions are not by chance or irrelevant. They arise in the context of the research situation and are informed by past experience and future expectation. As such they are critical and, indeed, underpin research and consultancy, which are as much experiential and emotional as cerebral. If we attempt to cut out these aspects of research, then we undermine the process.

Experienced qualitative researchers have always accepted the importance of emotion in research. This is nothing new, but it needs greater emphasis. By openly acknowledging the importance of emotion, research becomes richer and closer to 'real life' situations. In this way it enables more relevant knowledge generation. A veteran qualitative researcher described to me her intuition that 'something important was happening in the group' as 'a bit like butterflies in my stomach… hard to describe. It's like I know emotionally, before I get the intellectual understanding. I'm on to something important and I have to just wait and be alert. Sometimes it's when things are confused or when, for no apparent reason, I get really interested in what is going on.'

Gordon and Langmaid (1988: 141) discuss the importance of confusion in understanding emotional issues. 'Confusion …always indicates that you are on the road to understanding something'. 'Staying with' confusion or other emotional upheavals, noting changes in mood in a research situation, exploring boredom or anger or conflict, learning to take our own emotional responses seriously and reflect on them will all encourage new depths and breadth of understanding. These are areas we need to hone. We need to explicitly teach 'moderating feeling' as part of commercial qualitative training. Equally, we need to be explicit with clients about what we are doing. 'I'm going to allow research participants to express their anger

about this, not try to suppress it. They need to be angry—
and then we will work through it.'

We use this ability, this way of knowing, instinctively—as
we do in the rest of life—but how often do we teach
young researchers to recognise these moments. We
are more likely to encourage them *not* to trust their
emotions, to dampen down such 'instincts'. Of course,
there are good reasons for this—emotional responses
which are undisciplined are like loose cannons. We tend
to associate emotion with excess, lack of rationality and
distortion. However, Damasio does stress the importance
of *well targeted and well deployed emotion*. We have been
very effective at learning how to discipline and focus our
rational minds, but less effective at developing strategies
for utilizing emotional energy. Ignoring emotional content
in research is not the solution.

SOME IMPLICATIONS FOR QUALITATIVE PRACTICE

What does all this mean for day to day qualitative practice? Moving from commercial qualitative research as it is currently practiced, towards an emergent inquiry approach involves a change in research practice.

Problem definition as integral to the research inquiry

Currently, the client typically 'hands' the problem to the researcher. But why are researchers not involved in defining the research problem? Is this not part of the process of iterative learning? In defining a research 'problem', we are partially determining the 'answer', in the sense that the 'problem' is socially constructed, as is the 'answer'. Problem definition is often given insufficient attention.

In a linear model, it is the client's job to define 'the research problem'—which is inevitably defined from the client's perspective. This may be fine when the project is straightforward. However, more and more research projects are complex, not least because there are multiple stake-holders involved, each with different agendas.

Process consultancy is an approach to organizational learning, developed by Edgar Schein and now widely adopted within major corporations. It is used during organizational change programmes, to develop corporate strategy or to facilitate team building. In process consultancy, problem exploration is regarded as an essential part of the consultation process. Problem definition is jointly defined by the consultant and clients (Schein, 1999: 18).

Equally, from an emergent inquiry perspective, problem definition is a natural part of the research process, not a separate exercise. Facilitated workshops with clients, including a relevant mix of stakeholders from within—and possibly outside—the organization, can be invaluable to enable exploration of issues from different perspectives and to help define the 'problem' to be addressed. However, this is crossing a boundary. Traditionally, clients are 'supposed' to be able to define the problem and know what steps they need to take to 'solve' it. But, as Edgar Schein points out, the client 'often does not know what she is looking for and indeed should not really be expected to know' (Schein:5). A change of role expectations between clients and researchers is needed before problem definition can be openly acknowledged as a valid research area and can be viewed as an essential part of emergent inquiry.

Adopting a 'process consultancy' model, the researcher would help the client to develop the learning, to disseminate it throughout the organization and to help employees put it to practical use. This might be in the role of 'consumer champion', ensuring that the project stays true to consumer needs, or as an anchor, to help clients question and draw out implications from particular strategies. These organizational roles draw on qualitative researchers' psychological and cultural experience, as well as their knowledge of specific markets or social contexts. Working with clients in this way, to draw out the implications of research knowledge and move this forward within the organization, should be a key part of emergent inquiry and of maximizing the benefits of the research.

Emergent analysis and interpretation

Traditionally, the analysis of research data both during and, especially after, fieldwork stages, has been an essential part of the qualitative research process. This is the stage when the researchers craft a cohesive, plausible and client-useful narrative from the messy, often partial and contradictory, 'raw data', and loosely formed hypotheses. What happens, in emergent inquiry and collaborative research, to these analysis and interpretation stages? As illustrated in an earlier example, if observation or participation *becomes* the research then, over time, the qualitative skill set will be eroded. Equally, if the research frame changes, then analysis and interpretation need to be incorporated in a manner which fits the new method, if we are to retain research quality.

The analysis and interpretation required in emergent inquiry needs to reflect the method of inquiry itself— as with traditional scientific research. Analysis and interpretation are integral to the research process and exist alongside it, interwoven with the process from instigation to completion. Currently, although most commercial researchers would regard analysis as on-going throughout the research process, they would also agree that analysis is prioritized at the end of the fieldwork stage. In emergent inquiry, analysis happens alongside the research process, as iterative learning.

Schon (1982: 43) argues that practitioners need to develop reflection-in-action, using past knowledge to inform the present. Legitimization of research therefore comes from analysis and interpretations *which are appropriate to the nature of the study*. Training researchers and clients in reflective skills and personal awareness is therefore essential grounding for the effective practice of emergent inquiry. The personal qualities of the researcher, developed through formal training and mentorship, assumes great importance. Training of new commercial qualitative researchers in appropriate skill is a priority.

Living life as inquiry

When people say that we need to integrate research from different sources, they usually mean that qualitative research needs to be married up with other 'respectable' forms of research input, such as desk research, data bases, quantitative research. I'm not dismissing these research inputs, but it is not what I mean here.

Within our culture, we have grown to view 'research' as something separate from 'life' when, in practice, we experience the day to day world through 'research'; we observe, make meaning, create connections, test hypotheses, experiment. I am suggesting that we need to consciously lower the barrier between 'research' and 'life'. Instead of 'research' being corralled into research *method*, the researcher—along with everyone else involved in the project—would keep their eyes and ears open and gather clues and inspiration wherever it is to be found. This is already happening in much research, for example, with the increased emphasis on ethnographic techniques and pre and post focus group activities for research participants, but I think we need to go further, so research itself is less about research method and more about research as a way of understanding and thinking (in which method is a useful tool).

This perspective assumes that research fodder is all around, if we can recognise it, connect with it and allow it to feed our knowledge generation. The more we can get out of role—client, employee, researcher, consumer—whilst still using the experiences, knowledge, skills and rigour that are associated with that role, the more connections we can make and, potentially, the more creative and useful the outcome. Judi Marshall, an academic researcher, describes this process of 'living life as inquiry' as:

> By living life as inquiry I mean a range of beliefs, strategies and ways of behaving which encourage me to treat little as fixed, finished, clear-cut. Rather, I have an image of living continually in process, adjusting, seeing what emerges, bringing these things into question…In this integrated life, in which research is not separate or bounded, I _must_ hold an attitude of continuing inquiry, as I seek to live with integrity, believing in multiple perspectives rather than one truth …not separating off academic knowing from the rest of my activity. (Marshall, 1999: 155-171)

Research can be a process of on-going inquiry which is not bounded by the separation of reason and emotion, which is open to ideas wherever they emerge, which is a broader process than just 'research' by its traditional definition and which allows for a range of different perspectives.

How is 'Emergent Inquiry' legitimized?

Within everyday life, the notion of 'objective science' is good enough as a working model for understanding the mechanics of the world. However, it does not adequately account for creativity, for art, for pushing back frontiers. In these areas research can never be 'objective', as any pure scientist will tell us. 'Objectivity' always exists within a particular world view; the invisible web of rules, beliefs, assumptions that define our world. And therefore, is always relative. 'Truths' that we hold as sacred today, will be debunked in the future.

New scientific hypotheses and theories are the result of curiosity, engagement, inspired guesswork, a linking together of previously unconnected assumptions or empirical observations. These are always creative acts in which the scientist is a key player. It is only later, when the theory becomes established that the scientist distances him/herself from the discovery and it becomes a 'fact', supposedly independent of its creator. New understanding always start with this curiosity and engagement; connecting the previously unconnected. This is the 'edge of chaos', where true inquiry and original thinking is generated and it is where much qualitative research practice is located. But if we can't label it as 'the truth', how do we know if it has any worth at all?

A complementary way of understanding research discipline and rigour

From the perspective of 'Emergent Inquiry', knowledge that is generated in a research project is a form of truth which has been jointly developed by those involved in the project, at that time, within the context they are working in. If we assume that this 'truth' or knowledge is socially constructed, then the evaluation of its worth or legitimacy must also be socially constructed. We cannot sensibly apply measures of evaluation which derive from one approach to understanding knowledge (that it is discovered, objective, a 'thing')—to an approach which has arisen from an alternative understanding of knowledge (that it is never 'objective', that it is created, that it is an on-going, changing process).

However, although I would disagree with the idea of 'objective truth', I would argue that it is possible to achieve a relative 'truth'; a partial, situated, contextualized 'truth', which is socially constructed within a particular group or culture. If 'legitimization' of qualitative research knowledge cannot come from traditional means of validity and reliability used in classical science, then we must find it in a process of personal and shared reflection, reflexivity, analysis and questioning of the underlying basis for assumptions and theory. Ultimately legitimization comes from an assessment by all those involved, about whether the knowledge makes sense and makes a useful contribution to on-going understanding and decision making. Pragmatically, this involves the evaluation of plausibility and degree of fit with existing knowledge from other sources—though of course this needs to be treated with caution, to make sure that it does not simply reinforce the known. This process requires further reflection, questioning etc. In fact, the processes of reflection and analysis are on-going throughout the whole research process, not applied in retrospect. They are integral to the research inquiry.

However, 'reality' is the final judge. The world does not tolerate all understandings equally and the meaning we jointly create has to make sense within an existing socially constructed reality—and plausibly start to shift that 'reality'.

This process of legitimization is difficult to describe because, although it clearly involves a good deal of intellectual engagement, it also involves feeling, intuition, instincts, physiological reactions, many of which are tacit and difficult to pin down—although this does not make them any the less real. Legitimization, therefore, is partly a 'felt' process—a shared 'knowing' that what is emerging is 'right'. John Shotter and Donald Schon have gone some way towards describing this process, in 'action research' and 'reflection-in-action', as described in the previous section, but it is essentially experiential. As such, it is not easily communicated by words alone. In a sense, the 'teacher-pupil' relationship or apprenticeship—the process by which qualitative research was taught or 'passed on' in the '70s and early '80s (and to an extent is still taught) naturally lent itself to this way of learning.

In saying all of this, I am not claiming that research structure is redundant or that careful attention to samples, discussion guides, recruitment and presentation charts are not important. I am saying that these are hygiene factors. We need to go beyond them. And we need to take research knowledge laterally, not literally; developing the implications and connections rather than implementing the 'findings'. We need to absorb structure as a discipline and then move on—just as we might write a discussion guide to focus our attention or to reassure the client, but then ignore it in the real life situation. I am not advocating sloppy research. Far from it. In a sense this approach requires far more discipline than a formulaic approach to research; it requires an on-going monitoring of, and reflecting on, the appropriateness of action, interpretation and direction.

What I am describing is a process of on-going inquiry and legitimization which requires our participation as 'whole people', engaging all our facilities and those of our clients, of consumers and of other stakeholder groups, in order to jointly create knowledge which is relevant to the needs of our complex world. Legitimization, at its most basic level, it is the extent to which participants, given their different interpretations and interests, feel that the inquiry process has helped their understanding of the defined problem and generated possible routes forward. However, legitimization does not necessarily imply agreement. Weick points out that, 'people may not share meaning, they do share experience' (Weick, 1995: 188). In fact, Weick suggests that arguing is a crucial source of sense-making and that ambiguity allows people to maintain the perceptions of agreement which is necessary to working relationships; potentially, novelty and innovation emerge in the ways in which conflicting perspectives are explored and argued out.

So how, in practice, do we develop the ability to assess the legitimacy of our work or that of other people? Commercial researchers have long understood the importance of intuition, of understanding the importance of trusting emotional responses. The excerpt below from Gordon and Langmaid's book, is particularly good at conveying the processes of reflexivity, emotional and sensory awareness, intellectual analysis which contribute towards legitimization.

> *Now that the tape is rolling, you will re-experience elements of the group at a very profound level. You may see a series of flashbacks or hear snippets of conversation or see 'important body language. Don't force it, let it run to its natural conclusion and then look back over it. What was going on? Were process factors or task defences at play? Is this straight talking we're hearing? What is the psychological climate like—the emotional atmosphere in the room? If you ever wonder whether you're experiencing a true version of what happened or not, become aware of your own body as you replay the tape. You'll find yourself re-experiencing the postures, facial expressions, heart-rate, eye movements and so on of that time back there when you were in the group.* (Gordon & Langmaid, 1988: 141)

I am aware that, by advocating emotional knowing as a way of legitimizing Emergent Inquiry, I can be seen as playing into the hands of those with a positivist predisposition; inviting the accusation that validation or legitimization by the researcher him/herself is no validation at all. I am laying myself open to the charge of 'flaunting irrationality' or 'blatant subjectivity'. But, if we are to heal the rift between 'mind' and 'body' which has dogged us since Descartes, then emotion and reason need to be united as equals. Damasio talks of 'well-targeted' and 'well deployed' emotions. This, of course, is very different to emotion as a 'loose cannon'—emotion, like reason, emerges in more or less fruitful ways.

IN SUMMARY

Commercial qualitative research has evolved, through practice, to fit client needs. One consequence of this evolution has been a move away from the classic scientific paradigm towards a socially constructed perspective. As a further development, in this essay, we have explored 'emergent inquiry' as a methodology which incorporates a social constructivist perspective within a process of iterative learning and reflection-in-action (Schon, 1982: 43).

Emergent inquiry is a way of thinking which encourages us to approach research with a different mind-set, one which is more receptive to new ideas, with a creative outlook, openness, a spirit of humility and with collaboration, on all sides. It is a process of participative inquiry which includes researchers, clients, customers and other interested parties. However, this does not mean it that it should not be rigorous and disciplined. On the contrary, an emergent approach requires an intrinsic discipline within each stage of the research process and we need to work at bringing classical science and emergence closer together.

> *Instead of the either—or oscillation between formal systematicity and creativity as fixed and static 'points of view', surely there is now a need in all of science to understand how, dynamically, we can move between them, and in so doing, dialogically or chiasmicly relate them in a meaningful relation with each other.* (Shotter, 2003)

Adopting an 'emergent inquiry' approach to research has implications for commercial qualitative research practice and, in particular, for the skill set required of commercial researchers. Strong emphasis on appropriate training is needed, to ensure that the rigour of commercial qualitative research is maintained.

Ultimately, within a commercial environment, emergent inquiry will be judged in terms of how it helps researchers to help clients, rather than by theories that explain or define it. To summarize, some key practical guidelines for emergent inquiry are as follows:

- Curiosity and openness to whatever emerges is the starting point; to be engaged with the problem (even if not the product area) is a pre-requisite. Creating possible solutions to problems is part of being human, as well as part of being a researcher.

- Defining the problem is part of the solution. If we accept 'the problem' on a platter, we do not have the opportunity to help shape it. However, if we accept that knowledge is created, then the 'problem' is also created through the way we choose which aspects we will pay attention to. Shaping the question is central to the task of research, *is* part of the research, in that the way the question is shaped will inevitably influence the outcome.

- Knowledge is constantly moving on, being re-created. It is never static. We can get a fix on a particular research problem or issue and create potential solutions—and this is generally good enough for our purposes—but it is always work in progress, it can never be the final solution.

- Our role is to facilitate the creation of knowledge, but we do not have sole responsibility for delivering 'an answer' to the client. It is important that the thinking, guidance, potential routes that we offer can be further worked on with our clients. This ensures the greatest possible input to decision making. Working on potential directions together, as part of the feedback session, at separate sessions afterwards or as an on-going 'consumer champion' are fruitful way of bringing knowledge together, developing it and exploring implications

- We are participants in the research process, not observers. As such our emotional responses in the research situation—as well as those of research participants, clients and other stakeholders are just as valid as our rational responses—in fact they are invaluable—in understanding and constructing possible outcomes of the research.

- As researchers, starting from a position of ignorance is OK. We may be experts in research approaches, but we are probably not experts in the subject matter we are focussing on or in creating solutions to the particular problem we are exploring. In relation to our clients we often adopt the role of the 'supplier' or 'expert'. In either roles, the power balance with the client can get stuck and the possibilities for interaction and development of ideas become stilted. The roles define the interaction; clients feel they own the research question, researchers feel that they should not challenge the client. By putting roles to one side, the researcher would be able, as Schein puts it, 'to access her areas of ignorance' without fear of appearing stupid. Equally, the client could accept the questioning as part of the research process, rather than a challenge to his or her authority.

- Role boundaries are a distraction to learning. It is useful to start from the assumption that everyone has more to contribute to knowledge building than they consciously know, or that we can usually access. By developing an egalitarian culture, the contribution people can make is much broader, encompassing the personal as well as the professional—what Damasio describes as 'whole body' knowing.

- New and creative thinking occurs when we are outside our comfort zone, when we encourage diversity; it is our job to encourage people to think the unthinkable, do the undoable. This has implications for group structure. There are good reasons for having homogenous groups. There are also good reasons, on occasion, for doing the opposite.

- Commercial qualitative research has become wedded to the small group. In psycho-dynamic circles, the large group is a well used format and the differences between small and large groups has been written about extensively (Stacey, 2003; Dalal, 2002). Psychological research suggests that large groups (20+) often provoke deeper, more primitive feelings. They are more emotionally charged. Working with large groups has parallels with working within organizations and often triggers off patterns which replicate organizational conflict and problem resolution. Equally, they may replicate viral marketing patterns or the development of cult brands. It is time, as qualitative researchers, that we explored the possibilities of large groups for our practice.

REFERENCES AND BIBLIOGRAPHY

Alvesson, M. and Skoldberg, K. (2000). *Reflexive Methodology: New Vistas for Qualitative Research*, ISBN 9780803977068.

Bayley, G. (2006). "How do I know what I think until I hear what I say? Is it time to tear up the discussion guide?" Proceedings from the QRCA Conference, Atlanta.

Burr, V. (1995). *An Introduction to Social Constructionism*, ISBN 9780415104050.

Cassell, C., Johnson, P., Symon, G. and Bishop, V. (2005). *Benchmarking Good Practice in Qualitative Management Research*, ESRC Publication, http://shef. ac.uk/bgpinqmr/.

Chandler, J. and Owen, M. (2002). *Developing Brands with Qualitative Market Research*, ISBN 9781412903967.

Damasio, A. (2000). *The Feeling of What Happens*, ISBN 9780099288763.

Earls, M. (2002). *Welcome to the Creative Age: Bananas, Business and the Death of Marketing*, ISBN 9780470844991.

Ereaut, G. (2002). *Analysis and Interpretation in Qualitative Market Research*, ISBN 9781412903974.

Frongillo, E.A., Chowdhury, N., Ekstrom, E.-C. and Naved, R.T. (2003). "Understanding the experience of household food insecurity in rural Bangladesh leads to a measure different from that used in other countries," *Journal of Nutrition*, ISSN 0022-3166,133(December): 4158-62.

Goulding, C. (ed.) (2002). *Grounded Theory: A Practical Guide for Management, Business and Market Researchers*, ISBN 9780761966821.

Gordon, W. (1999). *Goodthinking: A Guide to Qualitative Research*, ISBN 9781841160306.

Gordon, W. and Langmaid, R. (1988). *Qualitative Market Research: A Practitioner and Buyers Guide*, ISBN 9780566051159.

Griffin, D. (2002). *The Emergence of Leadership: Linking Self Organization and Ethics*, ISBN 9780415249171.

Gummesson, E. (2000). *Qualitative Methods in Management Research*, ISBN 9780761920144.

Holmes, C. and Keegan, S. (1983). "Current and developing creative research methods in new product development," paper given at the MRS Annual Conference, Brighton.

Keegan, S. (2009a). *Qualitative Research: Good Decision Making through Understanding People, Cultures and Markets*, ISBN 9780749454647.

Keegan, S. (2009b). "'Emergent Inquiry': A Practitioner's reflections on the development of qualitative research," *Qualitative Market Research: An International Journal*, ISSN 1352-2752, 12(2): 234-248.

Keegan, S. (2008). *Re-defining Qualitative Research within a Business Context,* ISBN 9783836474177.

Keegan, S. (2006). "Emerging from the cocoon of science," *The Psychologist Magazine,* 19(1), http://www.thepsychologist.org.uk/.

Keegan, S. (2005). "Emergent inquiry: The 'new' qualitative research," paper given at the AQR/QRCA Bi-Annual Conference, Dublin.

Keegan, S. (2003). "A Lion in the Mist: Why does qualitative research keep underselling its strategic potential?" Paper given at the AQR/QRCA Bi-Annual Conference, Lisbon.

Kemmis, S. and McTaggart, R. (2000). "Participatory action research" in N.K. Denzin and Y.S. Lincoln (eds.), *The Sage Handbook of Qualitative Research,* 2nd Ed., ISBN 9780761927570.

Langmaid, R. and Andrews, M. (2003). *Breakthrough Zone: Harnessing Consumer Creativity for Business Innovation,* ISBN 9780470855393.

Lewin, R. (2001). *Complexity: Life At the Edge of Chaos,* ISBN 9780753812709.

Marshall, J. (1999). "Living life as inquiry," *Systemic Practice and Action Research,* ISSN 1094-429X, 12:155-171

Mead, G.H. (1962). *Mind, Self and Society: From the Standpoint of a Social Behaviorist,* ISBN 9780226516684.

Parnes, S.J. (1967). *Creative Behavior Guidebook,* ISBN 9780684413952.

Polanyi, M. (1962). *Personal Knowledge: Towards a Post-Critical Philosophy,* ISBN 9780226672885.

Puccio, G.J., Firestien, R.L., Coyle, C. and Masucci, C. (2006). "A review of the effectiveness of CPS training: A focus on workplace issues," *Creativity and Innovation Management,* ISSN 1077-2901, 15(1): 19-33.

Schein, E. (1999). *Process Consultation Revisited: Building the Helping Relationship,* ISBN 9780201345964.

Schon, D.A. (1983) *The Reflective Practitioner: How Professionals Think In Action,* ISBN 9780465068784.

Seel, R. (2000). "Culture and complexity: New insights on organizational change," *Organizations and People,* ISSN 1350-6269, 7(2): 82-84.

Senge, P. (1990). *The Fifth Disciple,* ISBN 9780712656870.

Shaw, P. (2002). *Changing Conversations in Organizations: A Complexity Approach to Change,* ISBN 9780415249140.

Shotter, J. (2003). "Participatory action research: A finished, classical science or a research science?" Unpublished article.

Shotter, J. (1993). *Conversational Realities: Constructing Life through Language,* ISBN 9780803989337.

Spencer, L., Ritchie, J., Lewis, J. and Dillon, L. (2003). *Quality in Qualitative Evaluation: A Framework for Assessing*

Research Evidence, ISBN 07115044658.

Stacey, R. (2011). *Personal communication.*

Stacey, R. (2001). *Complex Responsive Processes in Organizations,* ISBN 9780415249195.

Stacey, R. (1996). *Strategic Management and Organizational Dynamics,* ISBN 9780273642121.

Stacey, R., Griffin, D. and Shaw, P. (2000). *Complexity and Management: Fad or Radical Challenge to Systems Thinking,* ISBN 9780415247610.

Toffler, A. (1970). *Future Shock,* ISBN 9780330028615.

Valentine, V. (2002). "Repositioning research: A new MR language model," *International Journal of Market Research,* ISSN 1470-7853, Quarter 2.

Walker, C. and Foote, M. (2000). "Emergent inquiry: Using children's literature to ask hard questions about gender bias," *Childhood Education,* ISSN 0009-4056, 76(2): 88-91.

Watts, A. (1954). *The Wisdom of Insecurity,* ISBN 9780091210717.

Watts, A. (1969). *The Book on the Taboo Against Knowing Who You Are,* ISBN 9780679723004.

Weick, K.E. (1995). *Sensemaking in Organizations,* ISBN 9780803971776.

Lightning Source UK Ltd.
Milton Keynes UK
UKOW021042161011

180400UK00008B/1/P